Constructing a Perpendicular to a Line

Problem: *Construct a perpendicular to $\overleftrightarrow{XY}$ which passes through Y.*

X •————————————• Y

Solution:

1. Draw a long line through X and Y.

2. Draw two small arcs with center Y and radius $\overline{XY}$ which intersect the line.

3. Is one of the points of intersection X? _ _ _ _ _ _

   Label the other point of intersection Z.

4. Draw a large arc with center X and radius greater than $\overline{XY}$.

5. Use a congruent radius to draw a large arc with center Z.
   Make the arcs intersect in two points.

6. Label the points of intersection U and V.

7. Draw $\overleftrightarrow{UV}$.

8. Does $\overleftrightarrow{UV}$ pass through Y? _ _ _ _ _ _

9. Does $\overleftrightarrow{UV}$ bisect segment $\overline{XZ}$? _ _ _ _ _ _

   Is $\overleftrightarrow{UV}$ perpendicular to $\overleftrightarrow{XZ}$? _ _ _ _ _ _

   Is $\overleftrightarrow{UV}$ the perpendicular bisector of segment $\overline{XZ}$? _ _ _ _ _ _

Problem: *Construct a perpendicular to $\overleftrightarrow{AB}$ which passes through A.*

A •————————————————• B

Solution:

1. Draw a long line through A and B.

2. Draw two small arcs with center A and radius $\overline{AB}$ which intersect the line.

3. Is one of the points of intersection B? _ _ _ _ _
   Label the other point of intersection C.

4. Construct the perpendicular bisector of segment $\overline{CB}$.

5. Does the perpendicular bisector pass through A? _ _ _ _ _

Key to

Geometry®

5
Student
Workbook

Squares and Rectangles

By Newton Hawley and Patrick Suppes
Revised by George Gearhart and Peter Rasmussen

Name _____   Class _____

KEY TO GEOMETRY
Book 5: Squares and Rectangles
TABLE OF CONTENTS

TO THE STUDENT:

These books will help you to discover for yourself many important relationships of geometry. Your tools will be the same as those used by the Greek mathematicians more than 2000 years ago. These tools are a **compass** and a **straightedge**. In addition, you will need a **sharpened pencil**. The lessons that follow will help you make drawings from which you may learn the most.

The answer books show **one** way the pages may be completed correctly. It is possible that your work is correct even though it is different. If your answer differs, re-read the instructions to make sure you followed them step by step. If you did, you are probably correct.

Cover art by Howard Coale.

Euclid is the best known geometer in all of history. He taught at the Greek school of mathematics in Alexandria, Egypt. Euclid also wrote the Elements, a series of 13 books written on papyrus rolls which summed up all the geometry that was known in his time. Almost everything that you will study in Key to Geometry and in other school geometry books was contained in the Elements, written about 300 B.C., more than 2200 years ago.

On the cover of this booklet Euclid draws with a compass and straightedge. These two tools of geometry are sometimes called the "Euclidean tools" because they were very important in Euclid's work.

® Key to Fractions, Key to Decimals, Key to Percents, Key to Algebra, Key to Geometry, Key to Measurement, and Key to Metric Measurement are registered trademarks of Key Curriculum Press.
Published by Key Curriculum Press, 1150 65th Street, Emeryville, CA 94608
Printed in the United States of America   32 31   08 07 06   ISBN 0-913684-75-9

Problem: *Construct a perpendicular to $\overleftrightarrow{PQ}$ which passes through P.*

Solution:

1. Extend line $\overleftrightarrow{PQ}$.

2. Construct a perpendicular to the line through point P.

3. Construct a perpendicular to $\overleftrightarrow{RS}$ which passes through S.

R ————————————————— S

A

B

1. Construct perpendiculars to $\overline{AB}$ at A and at B.

2. Construct a perpendicular to $\overline{PQ}$ at P.

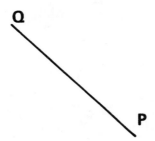

Q

P

3. Label as R a point on this perpendicular.

4. Construct a perpendicular to $\overleftrightarrow{PR}$ at R.

Review

1. Bisect the angle.

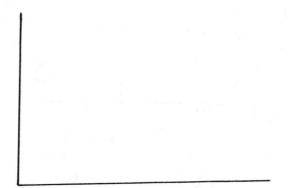

2. Which angle is larger? _ _ _ _ _

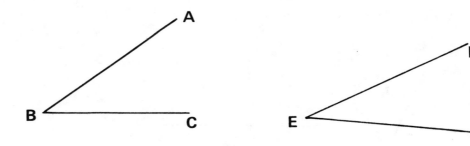

3. Construct a segment congruent to $\overline{AB}$ on the given line.

1. Construct the perpendicular bisector of the given segment.

———————————

2. Construct a perpendicular to each line through the given point.

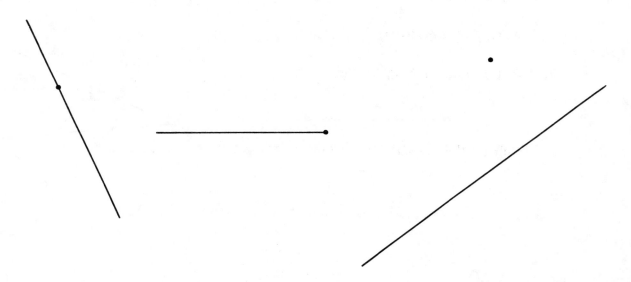

3. Is ABC a right angle? _ _ _ _ _

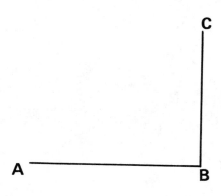

Squares

ABCD is a <u>square.</u>

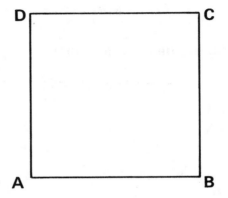

1. Compare the sides of the square.

   Are they all congruent? _ _ _ _ _

2. Are the angles all right angles? _ _ _ _ _

3. Which of the quadrilaterals below is a square? _ _ _ _ _

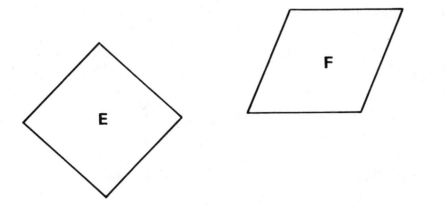

A square is a quadrilateral which has all its sides congruent and all its angles right angles.

Which of the figures below is a square? _ _ _ _ _

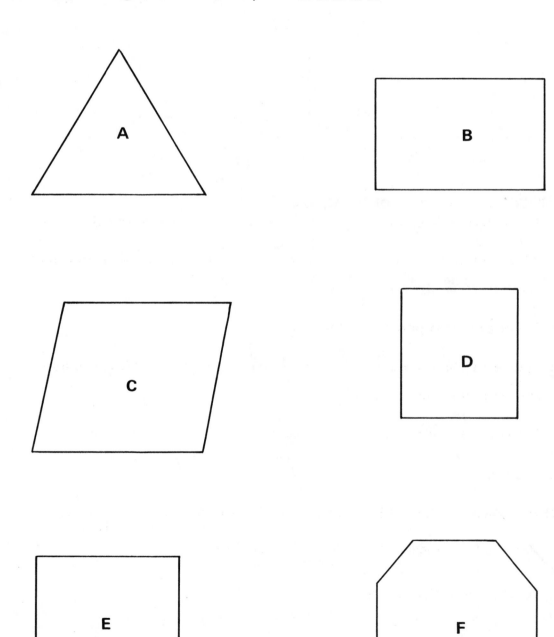

Constructing Squares

Problem: *Construct a square.*

```
————————————•————————————————————•————————————
             A                     B
```

Solution:

1. Draw perpendiculars to segment $\overline{AB}$ through points A and B.

2. Above the line draw an arc with center A and radius $\overline{AB}$ to intersect the perpendicular through A.

3. Label as C this point of intersection.

4. Above the line draw an arc with center B and radius $\overline{AB}$ to intersect the perpendicular through B.

5. Label as D this point of intersection.

6. Draw $\overline{CD}$.

7. Are the four sides $\overline{AB}$, $\overline{BD}$, $\overline{CD}$, and $\overline{AC}$ all congruent? _ _ _ _ _

   Are the four angles of ABDC all right angles? _ _ _ _ _

8. What kind of figure is ABDC? _ _ _ _ _ _ _ _ _ _ _ _ _ _ _ _ _

Problem: *Construct a square with a given segment as a side.*

P ——————————————— Q

Solution:

1. Construct a perpendicular above $\overleftrightarrow{PQ}$ which passes through P.

2. Construct a perpendicular above $\overleftrightarrow{PQ}$ which passes through Q.

3. Draw an arc with center P and radius congruent to $\overline{PQ}$ which intersects the perpendicular through P.

4. Label the intersection X.

5. Draw an arc with center Q and radius congruent to $\overline{PQ}$ which intersects the perpendicular through Q.

6. Label the intersection Y.

7. Draw $\overline{XY}$.

8. Is PQYX a square? _ _ _ _ _

1. Construct a square with the given segment as a side.

_____

Are all the angles of the square right angles? _ _ _ _ _

2. Construct another square.

ABYX is a square.

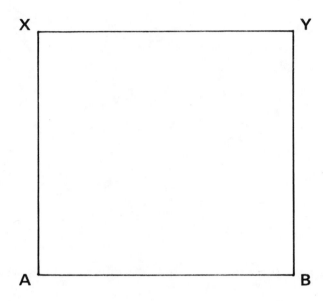

1. Bisect side $\overline{AB}$ and label the midpoint C.

2. Bisect side $\overline{AX}$ and label the midpoint U.

3. Bisect side $\overline{BY}$ and label the midpoint V.

4. Bisect side $\overline{XY}$ and label the midpoint Z.

5. Draw $\overline{CZ}$ and $\overline{UV}$.

6. How many squares do you see? _ _ _ _ _

X —————————————— Y

1. Construct a perpendicular to $\overleftrightarrow{XY}$ which passes through X.

2. Above X on the perpendicular construct a segment congruent to $\overline{XY}$ and label as Z its other endpoint.

3. Construct a perpendicular to $\overleftrightarrow{XY}$ which passes through Y.

4. Above Y on this perpendicular construct a segment congruent to $\overline{XY}$ and label as W its other endpoint.

5. Draw $\overline{ZW}$.

6. Is this quadrilateral a square? _ _ _ _ _

Review

1. Compare the angles.

   Are they congruent? _ _ _ _ _

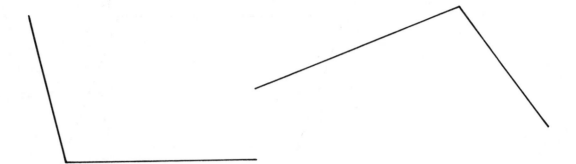

2. Construct a perpendicular to the line through the given point.

3. Bisect the angle.

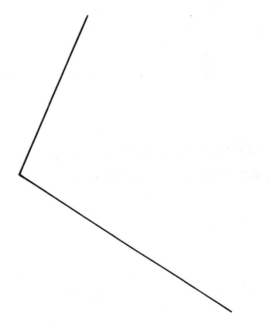

Rectangles

A <u>rectangle</u> is a quadrilateral with all its angles right angles.

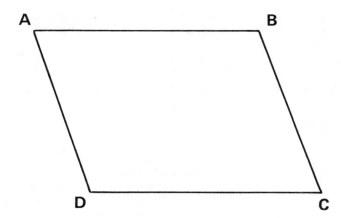

1. Is ABCD a rectangle? _ _ _ _ _

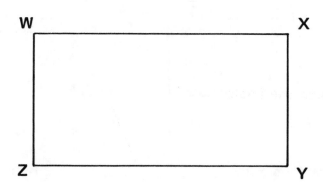

2. Check: Is WXYZ a rectangle? _ _ _ _ _
   (Do not guess. Use your compass.)

ABCD is a rectangle.

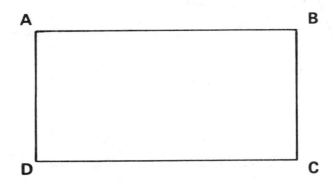

1. Are all four sides congruent? _ _ _ _ _

2. Side $\overline{AD}$ is congruent to side _ _ _ _ _ .

   Side $\overline{AB}$ is congruent to side _ _ _ _ _ .

4. Check: Is the angle at C a right angle? _ _ _ _ _

5. A rectangle has _ _ _ _ _ _ _ _ _ _ right angles.
   (a) one   (c) three
   (b) two   (d) four

1. Which of the figures below is a rectangle? _ _ _ _ _

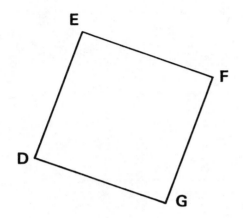
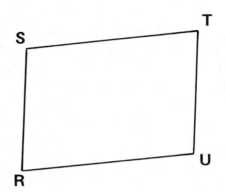

2. Is WXYZ a rectangle? _ _ _ _ _

   Why? _

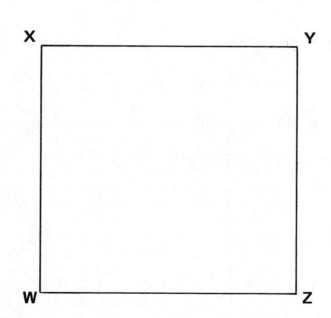

3. Is WXYZ also a square? _ _ _ _ _

1. Does every square have all right angles? _ _ _ _ _

   Is every square a rectangle? _ _ _ _ _

2. Is the quadrilateral ABCD a rectangle? _ _ _ _ _

   How do you know? _

3. Is the quadrilateral ABCD a square? _ _ _ _ _

4. Are there some rectangles which are not squares? _ _ _ _ _

1. Are all squares rectangles? _ _ _ _ _

2. Is every rectangle a square? _ _ _ _ _

3. Which of the figures below is a rectangle? _ _ _ _ _
   (Don't guess. Use your compass.)

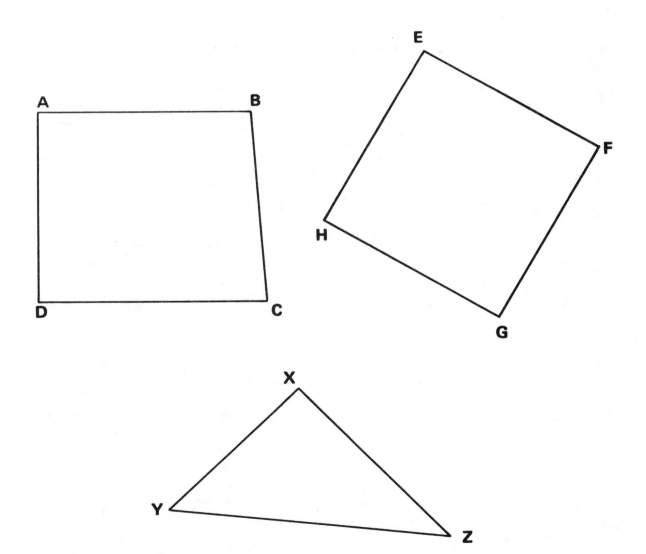

Review

1. Are the angles congruent? _ _ _ _ _

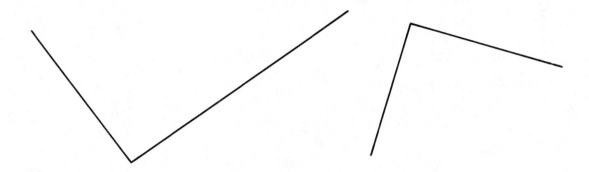

2. Construct the perpendicular to each line through the given point.

Constructing Rectangles

Problem: *Construct a rectangle with a given base.*

P ─────────────────────────── Q

Solution:

1. Extend $\overleftrightarrow{PQ}$ to the left.

2. Draw arcs with center P and the same radius which intersect the extended line in two points. Label the points of intersection U and V.

3. Draw an arc with center U above P.

4. Use the same radius to draw an arc with center V above P.
   Make these two arcs intersect. Label their point of intersection W.

5. Draw $\overleftrightarrow{PW}$.

   Is PW perpendicular to $\overleftrightarrow{PQ}$? _ _ _ _ _

6. Choose a point on $\overleftrightarrow{PW}$ and label the point as A.

7. Construct a perpendicular to $\overleftrightarrow{PQ}$ through Q.

8. Draw an arc with center Q and radius congruent to $\overline{PA}$ above Q.

9. Label as B the intersection of this arc with the perpendicular.

10. Draw $\overline{AB}$.

11. Is ABQP a rectangle? _ _ _ _ _

Problem: *Construct a rectangle with a given base.*

P ———————————————————————— Q

Solution:

1. Construct a perpendicular to $\overleftrightarrow{PQ}$ through P.

2. Construct a perpendicular to $\overleftrightarrow{PQ}$ through Q.

3. Choose a point on the perpendicular through P and label it A.

4. Draw an arc with center Q and radius congruent to $\overline{PA}$ which intersects the perpendicular through Q.

5. Label the intersection as B.

6. Draw $\overline{AB}$.

7. Is ABQP a rectangle? _ _ _ _ _

   How do you know? _

Problem: *Construct a rectangle on the segment $\overline{PQ}$ with one side congruent to segment $\overline{PQ}$, and one side congruent to segment $\overline{AB}$.*

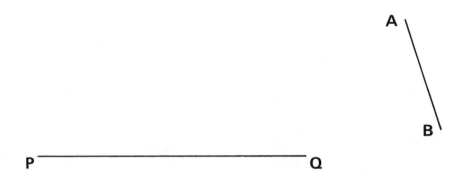

Solution:

1. Construct a perpendicular to $\overleftrightarrow{PQ}$ through P.

2. Draw an arc with center P and radius congruent to $\overline{AB}$ which intersects the perpendicular above P.

3. Label the intersection L.

4. Construct a perpendicular to $\overleftrightarrow{PQ}$ through Q.

5. Draw an arc with center Q and radius congruent to $\overline{AB}$ which intersects the perpendicular above Q.

6. Label the intersection M.

7. Draw $\overline{LM}$.

Problem: *Construct a rectangle on a given line with one side congruent to segment $\overline{XY}$ and one side congruent to segment $\overline{AB}$.*

Solution:

1. On the given line construct a segment congruent to $\overline{XY}$. Label its endpoints as P and Q.

2. Construct perpendiculars to the line through P and through Q.

3. Label points L and M on the perpendiculars so that segment $\overline{LP}$ is congruent to segment $\overline{AB}$ and segment $\overline{MQ}$ is congruent to segment $\overline{AB}$.

4. Draw $\overline{LM}$.

Problem: *Construct a rectangle with all four sides congruent to a given segment.*

P ——————————————— Q

Solution:

1. Construct perpendiculars to $\overleftrightarrow{PQ}$ through P and through Q.

2. Label points L and M on these perpendiculars so that segment $\overline{LP}$ is congruent to segment $\overline{PQ}$ and segment $\overline{MQ}$ is congruent to segment $\overline{PQ}$.

3. Draw $\overline{LM}$.

4. Is LMQP a rectangle? _ _ _ _ _

   How do you know? _

   _

5. Is LMQP a square? _ _ _ _ _

   How do you know? _

   _

1. Construct a rectangle with one side congruent to segment $\overline{AB}$ and one side congruent to segment $\overline{CD}$.

A _____ B

2. Construct a square with $\overline{PQ}$ as one side.

P _____ Q

Review

1. Construct the perpendicular bisector of the given segment.

2. Is the angle a right angle? _ _ _ _ _

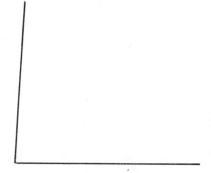

3. Bisect the given angles.

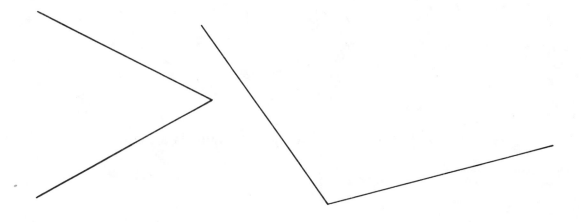

The Diameter of a Circle

A line segment passing through the center of a circle, which has both its endpoints on the circle, is a <u>diameter</u> of the circle.

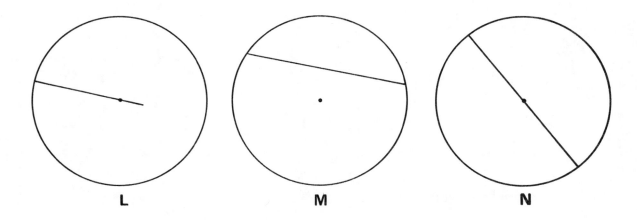

L   M   N

1. Which of the circles above has a diameter drawn? _ _ _ _ _

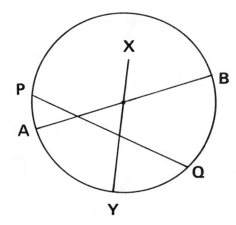

2. Which of the line segments in the figure is a diameter of the circle? _ _ _ _ _

1. Draw three diameters in each of the circles.

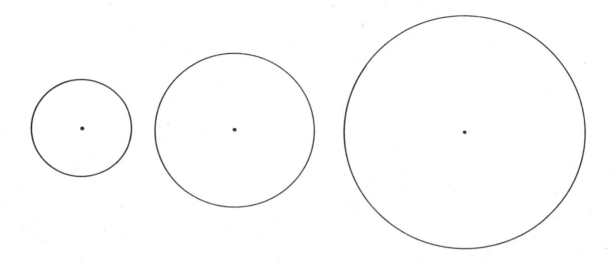

2. Draw a diameter of the circle below and label its endpoints P and Q.

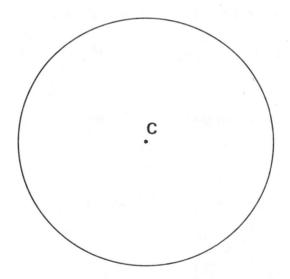

3. The point C is the center of the circle.

   Is C also the midpoint of diameter $\overline{PQ}$? _ _ _ _ _

4. Construct a perpendicular to $\overleftrightarrow{PQ}$ which passes through C.

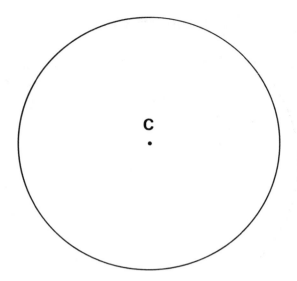

1. Draw a diameter of the circle and label its endpoints as P and Q.

2. Construct a perpendicular to $\overleftrightarrow{PQ}$ which passes through C. Make the perpendicular intersect the circle in two points.

3. Label as A and B the two points in which this perpendicular intersects the circle.

4. Is the segment $\overline{AB}$ a diameter of the circle? _ _ _ _ _

5. Connect P and A.

6. Connect Q and A.

7. Is $\overleftrightarrow{PA}$ perpendicular to $\overleftrightarrow{PQ}$? _ _ _ _ _

   Is $\overleftrightarrow{PA}$ perpendicular to $\overleftrightarrow{AQ}$? _ _ _ _ _

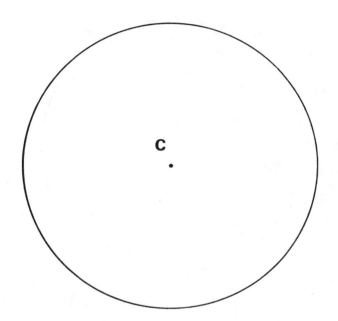

1. Draw a diameter of the circle and label its endpoints as N and P.

2. Construct a diameter of the circle which is perpendicular to $\overleftrightarrow{NP}$.

3. Label as S and T the endpoints of this new diameter.

4. Draw $\overline{NS}$, then draw $\overline{SP}$, then draw $\overline{PT}$, then draw $\overline{TN}$.

5. Check: Is $\overleftrightarrow{NS}$ perpendicular to $\overleftrightarrow{SP}$? _ _ _ _ _

6. Check: Is segment $\overline{NS}$ congruent to segment $\overline{SP}$? _ _ _ _ _

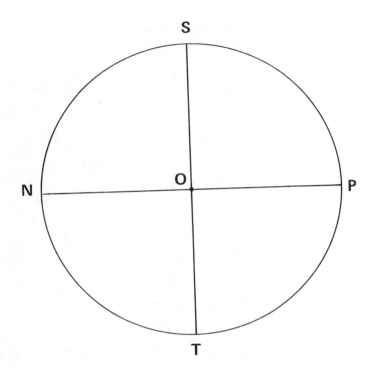

1. Check: Is diameter $\overline{NP}$ perpendicular to diameter $\overline{ST}$? _ _ _ _ _

   (Don't guess.)

2. Draw $\overline{NS}$, then draw $\overline{SP}$, then draw $\overline{PT}$, then draw $\overline{TN}$.

3. Check: Is $\overleftrightarrow{NS}$ perpendicular to $\overleftrightarrow{TN}$? _ _ _ _ _

4. Check: Is segment $\overline{NS}$ congruent to segment $\overline{TN}$? _ _ _ _ _

5. Is NSPT a square? _ _ _ _ _

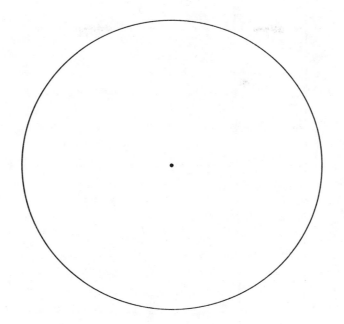

1. Draw a diameter of the circle and label its endpoints as A and B.

2. Draw another diameter of the circle and label its endpoints as C and D.

3. Draw figure ACBD.

4. What kind of figure is ACBD? _ _ _ _ _ _ _ _ _ _ _

The Diagonals of a Square

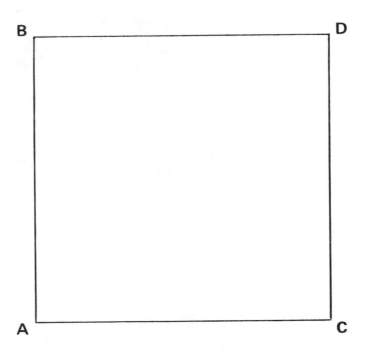

1. The figure ABDC is a square.

   (Check this statement.)

2. Draw $\overline{BC}$. The segment $\overline{BC}$ is a diagonal of the square.

3. Draw $\overline{AD}$. The segment $\overline{AD}$ is the other diagonal of the square.

4. Check: Is diagonal $\overline{BC}$ congruent to diagonal $\overline{AD}$? _ _ _ _ _

5. Check: Is $\overleftrightarrow{BC}$ perpendicular to $\overleftrightarrow{AD}$? _ _ _ _ _

6. How many diagonals does a square have? _ _ _ _ _

1. Construct a square with segment $\overline{PQ}$ as one side.

P ————————————————— Q

2. Bisect the angle at P.

3. Will the bisector pass through the opposite vertex? _ _ _ _ _

4. Draw the diagonals of the square ABCD.

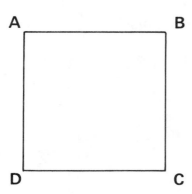

5. Fold your paper on $\overleftrightarrow{AC}$.

   Does $\overleftrightarrow{AC}$ bisect segment $\overline{BD}$? _ _ _ _ _

38

The Diagonals of a Rectangle

1. The figure WXZY is a rectangle.

   (Check this statement.)

2. Draw $\overline{WZ}$. The segment $\overline{WZ}$ is a diagonal of the rectangle.

3. Draw $\overline{XY}$. The segment $\overline{XY}$ is a diagonal of the rectangle.

4. Check: Is diagonal $\overline{XY}$ congruent to diagonal $\overline{WZ}$? _ _ _ _ _

5. Check: Is $\overleftrightarrow{XY}$ perpendicular to $\overleftrightarrow{WZ}$? _ _ _ _ _

6. Are the diagonals of a rectangle perpendicular? _ _ _ _ _

7. Are the diagonals of a square perpendicular? _ _ _ _ _

8. Are the diagonals of a rectangle congruent? _ _ _ _ _

9. Are the diagonals of a square congruent? _ _ _ _ _

1. The figure ABCD is a square.

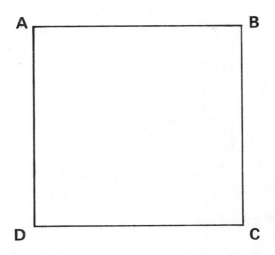

2. Draw diagonal $\overline{AC}$.

3. Construct the perpendicular bisector of diagonal $\overline{AC}$.

4. Does the perpendicular pass through B and D? _ _ _ _ _

5. PQRS is a rectangle.

   Draw $\overline{PR}$.

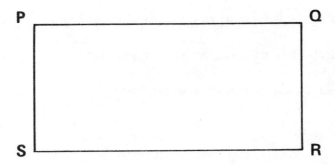

6. Construct the perpendicular bisector of diagonal $\overline{PR}$.

7. Does the perpendicular bisector pass through Q and S? _ _ _ _ _

1. Is ABCD a square? _ _ _ _ _

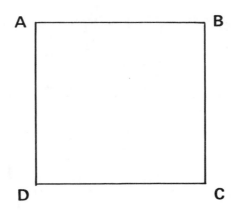

2. Draw $\overline{AC}$ and $\overline{BD}$.

   Label their intersection P.

3. Do you think you can draw a circle with center P which passes through

   points A, B, C, and D? _ _ _ _ _

4. Draw a circle with center P through point A and check your answer.

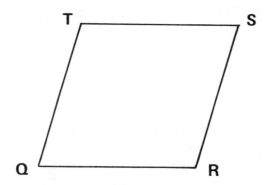

5. Is QRST a square? _ _ _ _ _

6. Draw $\overline{TR}$ and $\overline{SQ}$.

   Label their intersection O.

7. Can you draw a circle with center O which passes through points

   Q, R, S, and T? _ _ _ _ _

8. Draw a circle with center O through point Q and check your answer.

ABCD is a rectangle.

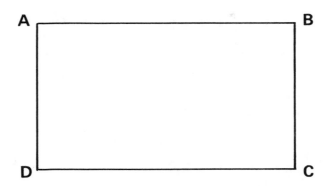

1. Draw $\overline{AC}$ and $\overline{BD}$.

   Label their intersection X.

2. Do you think you can draw a circle with center X passing through

   points A, B, C, and D? _ _ _ _ _

3. Draw a circle with center X through point A and check your answer.

4. Construct a circle which passes through vertices P, Q, R, and S of the

   square below.

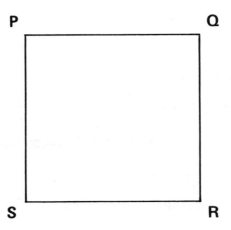

Review

1. Which angle is larger? _ _ _ _ _ _ _ _ _ _ _

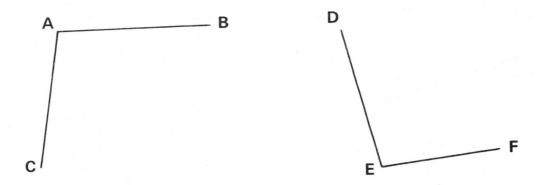

2. Check: Are the segments congruent? _ _ _ _ _ _

3. Construct a segment congruent to segment $\overline{AB}$ on the given line.

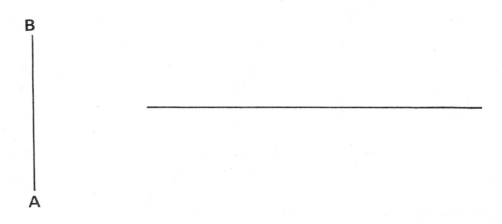

1. Construct perpendiculars to the given lines through the given points.

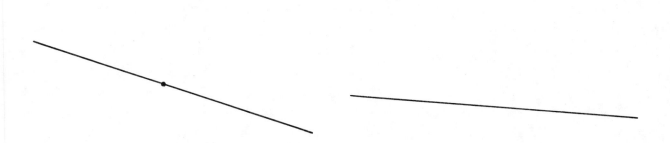

2. Bisect the angle.

3. Construct a square with the given segment as one side.

4. Draw the diagonals of the square.

Constructing a Segment Congruent to a Fraction of a Given Segment

1. Divide the given segment into two congruent parts.

   (Hint: Bisect the segment.)

   _____

2. Each part is _ _ _ _ _ _ _ _ _ _of the segment.

   (a) one <u>half</u>   (c) one <u>fourth</u>

   (b) one <u>third</u>   (d) one <u>eighth</u>

3. Divide the segment below into four congruent parts.

4. Each part is _ _ _ _ _ _ _ _ _ _of the segment.

   (a) one half   (c) one fourth

   (b) one third   (d) one eighth

Problem: *On a given line, construct a segment congruent to three halves of a given segment.*

A ————————————————— B

W ————————————————————————

Solution:

1. Bisect segment $\overline{AB}$.

   Label the midpoint M.

2. On the given line, construct a segment with endpoint W which is congruent to segment $\overline{AM}$. Label its other endpoint X.

3. To the right of X, construct a segment with endpoint X which is congruent to segment $\overline{AM}$. Label its other endpoint Y.

4. To the right of Y, construct a segment with endpoint Y which is congruent to segment $\overline{AM}$. Label its other endpoint Z.

5. Segment $\overline{WZ}$ is congruent to _ _ _ _ _ _ _ _ _ _ _of segment $\overline{AB}$.

    (a) one half   (c) one fourth

    (b) three halves   (d) three fourths

1. On the given line, construct a segment congruent to three halves of segment $\overline{CD}$.

C ———————————— D

————————————————————————————

2. Divide segment $\overline{PQ}$ into fourths.

P ———————————— Q

3. On the given line, construct a segment congruent to three fourths of segment $\overline{PQ}$.

————————————————————————————

Problem: *Construct a segment two and a half times as long as segment $\overline{AB}$.*

A ——————————— B

———————————————————————

Solution:

1. On the given line, construct a segment congruent to segment $\overline{AB}$. Label its endpoints M and N. (Put N on the right.)

2. To the right of N construct another segment congruent to segment $\overline{AB}$ with endpoint N. Label its other endpoint O.

3. Bisect segment $\overline{AB}$. Label its midpoint C.

4. To the right of O, construct a segment congruent to segment $\overline{AC}$ with endpoint O. Label the other endpoint P.

5. Segment $\overline{MP}$ is _ _ _ _ _ _ _ _ _ _ _ _ _ _ _ _ _ _ segment $\overline{AB}$.

(a) double

(c) two and a half times as long as

(b) triple

(d) one half as long as

1. On the given line, construct a segment two and a half times as long as segment $\overline{XY}$.

X ——————————— Y

———————————————————

Label the endpoints A and B.

2. On the line below, construct a segment congruent to five halves of segment $\overline{XY}$.

———————————————————

Label the endpoints C and D.

3. Segment $\overline{CD}$ is _ _ _ _ _ _ _ _ _ _ segment $\overline{AB}$.

   (a) shorter than   (c) longer than

   (b) congruent to

4. On the line below, construct a segment one and a half times as long as segment $\overline{XY}$.

———————————————————

Review

1. Check: Are the angles congruent? _ _ _ _ _

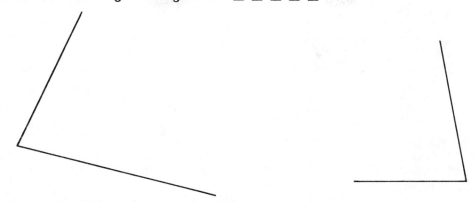

2. Construct a perpendicular to the line through the given point.

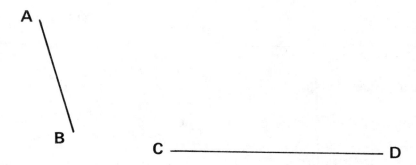

3. Construct a rectangle with one side congruent to segment $\overline{AB}$ and the other side congruent to segment $\overline{CD}$.

A

B

C ——————————— D

Congruent Polygons

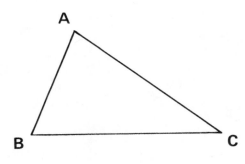

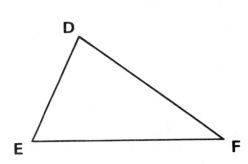

1. Trace triangle ABC.

2. Place your tracing on triangle DEF and try to make the triangles match.

3. If they can be made to match, triangle ABC is congruent to triangle DEF.

   Is triangle ABC congruent to triangle DEF? _ _ _ _ _

4. Point A <u>corresponds</u> to point _ _ _ _ _.

   Point B corresponds to point _ _ _ _ _.

   Point C corresponds to point _ _ _ _ _.

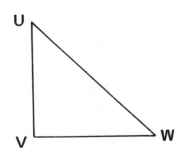

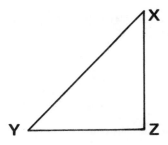

5. Trace triangle UVW.

6. Is triangle UVW congruent to triangle XYZ?_ _ _ _ _

7. Point U corresponds to point _ _ _ _ _.

   Point V corresponds to point _ _ _ _ _.

   Point W corresponds to point _ _ _ _ _.

1. Trace triangle PQR.

2. Place your tracing on triangle UST.

3. Can you make the triangles match exactly? _ _ _ _ _

4. Turn your tracing over.

5. Now can you make the triangles match? _ _ _ _ _

6. Is triangle PQR congruent to triangle UST? _ _ _ _ _

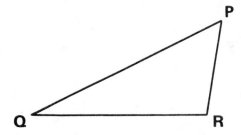
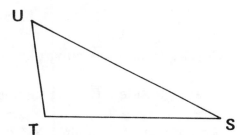

7. Is rectangle ABCD congruent to rectangle WXYZ? _ _ _ _ _

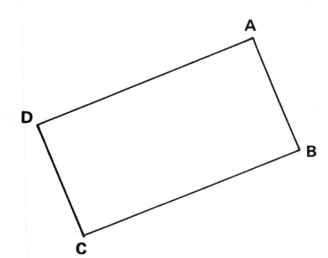
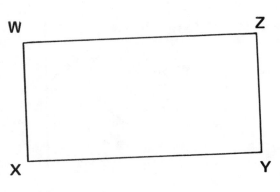

The given triangles are congruent.

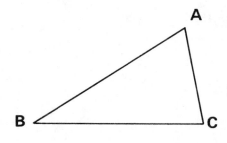
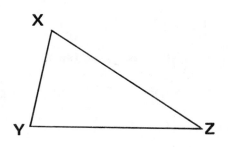

1. Check: Is side $\overline{AB}$ congruent to side $\overline{XZ}$? _ _ _ _ _

2. Check: Is side $\overline{AC}$ congruent to side $\overline{XY}$? _ _ _ _ _

3. Which side of triangle XYZ is congruent to side $\overline{BC}$? _ _ _ _ _

4. Check: Is angle ABC congruent to angle XZY? _ _ _ _ _

5. Which angle of triangle XYZ is congruent to angle BAC? _ _ _ _ _

6. Which angle of triangle XYZ is congruent to angle ACB? _ _ _ _ _

The given quadrilaterals are congruent.

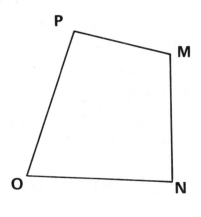

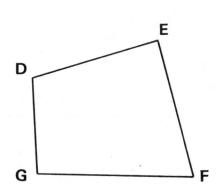

1. Side $\overline{GF}$ is congruent to side _ _ _ _ _.

   Side $\overline{GD}$ is congruent to side _ _ _ _ _.

   Side $\overline{DE}$ is congruent to side _ _ _ _ _.

   Side $\overline{EF}$ is congruent to side _ _ _ _ _.

2. Angle DGF is congruent to angle _ _ _ _ _.

   Angle GDE is congruent to angle _ _ _ _ _.

   Angle DEF is congruent to angle _ _ _ _ _.

   Angle EFG is congruent to angle _ _ _ _ _.

Practice Test

1. On the given line construct a segment congruent to three halves of segment $\overline{PQ}$.

P _____ Q

_____

2. Is ABCD a quadrilateral? _ _ _ _ _

   Is ABCD a rectangle? _ _ _ _ _

   Is ABCD a square? _ _ _ _ _

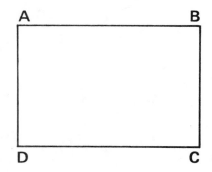 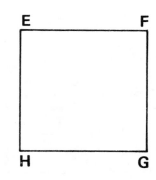 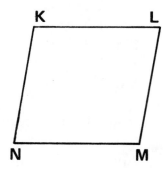

3. Is EFGH a quadrilateral? _ _ _ _ _

   Is EFGH a rectangle? _ _ _ _ _

   Is EFGH a square? _ _ _ _ _

4. Is KLMN a quadrilateral? _ _ _ _ _

   Is KLMN a rectangle? _ _ _ _ _

   Is KLMN a square? _ _ _ _ _

5. Are the given triangles congruent? _ _ _ _ _

6. Construct a rectangle with one side congruent to segment $\overline{AB}$ and one side congruent to segment $\overline{CD}$.

A _____ B

7. Segment _ _ _ _ _ is a diameter of the circle.

Segment _ _ _ _ _ is a radius of the circle.

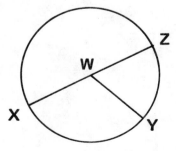

Key to Geometry®

Also Available

Key to Fractions®
Key to Decimals®
Key to Percents®
Key to Algebra®
Key to Measurement®
Key to Metric Measurement®

KEY CURRICULUM PRESS
Innovators in Mathematics Education

ISBN 0-913684-75-9